50 selected problems from past math competitions with detailed correction to get ready the Junior Mathematics Olympiad (or other math challenges)
Introductory level

Copyright © 2023 by Math's up
All rights reserved, including the right to reproduce this book or portions thereof in any form whatsoever

Math's up: mathsup314116@gmail.com

Copyright © 2023 by Math's up

All rights reserved.

No portion of this book may be reproduced in any form without written permission from the publisher or author, except as permitted by international copyright law.

This publication is designed to provide accurate and authoritative information in regard to the subject matter covered. It is sold with the understanding that neither the author nor the publisher is engaged in rendering legal, investment, accounting or other professional services. While the publisher and author have used their best efforts in preparing this book, they make no representations or warranties with respect to the accuracy or completeness of the contents of this book and specifically disclaim any implied warranties of merchantability or fitness for a particular purpose. No warranty may be created or extended by sales representatives or written sales materials. The advice and strategies contained herein may not be suitable for your situation. You should consult with a professional when appropriate. Neither the publisher nor the author shall be liable for any loss of profit or any other commercial damages, including but not limited to special, incidental, consequential, personal, or other damages.

1st edition 2023

Table of contents

Introduction 8

Problems 11

    Ex 1 (from UKMT 2021) 12

    Ex 2 (from OMB 2022) 12

    Ex 3 (from UKMT 2019) 12

    Ex 4 (from POFM 2015) 12

    Ex 5 (from UKMT 2021) 13

    Ex 6 (from AMC 2023) 13

    Ex 7 (from POFM 2016) 13

    Ex 8 (from PMO 2017) 14

    Ex 9 (from POFM 2019) 14

    Ex 10 (from UKMT 2021) 14

    Ex 11 (from SMO 2021) 14

    Ex 12 (from PMO 2009) 15

    Ex 13 (from UKMT 2022) 15

    Ex 14 (from POFM 2015) 15

    Ex 15 (from UKMT 2021) 15

    Ex 16 (from SMO 2022) 16

    Ex 17 (from OMB 2020) 16

    Ex 18 (from PMO 2016) 16

    Ex 19 (from CMO 2019) 17

    Ex 20 (from UKMT 2021) 17

    Ex 21 (from SMO 2020) 18

    Ex 22 (from AMC 2023) 18

Ex 23 (from POFM 2023) 18

Ex 24 (from UKMT 2021) 19

Ex 25 (from UKMT 2021) 19

Ex 26 (from SMO 2022) 19

Ex 27 (from PMO 2018) 19

Ex 28 (from OMS 2023) 19

Ex 29 (from AIME 2021) 20

Ex 30 (from POFM 2020) 20

Ex 31 (from Kangaroo 2022) 20

Ex 32 (from CJMO 2019) 21

Ex 33 (from AMC 2023) 21

Ex 34 (from SMO 2022) 21

Ex 35 (from UKMT 2021) 21

Ex 36 (from AMC 2022) 22

Ex 37 (from POFM 2014) 22

Ex 38 (from PMO 2023) 23

Ex 39 (from POFM 2022) 23

Ex 40 (from AIME2020) 23

Ex 41 (from SMO 2022) 23

Ex 42 (from Kangaroo 2015) 24

Ex 43 (from UKJMO 2015) 24

Ex 44 (from POFM 2017) 24

Ex 45 (from MacLaurin 2017) 24

Ex 46 (from AIME 2021) 25

Ex 47 (from RMC 2017) 25

Ex 48 (from PMO 2022) 25

Ex 49 (from CMO 2023) 25

Ex 50 (from AIME 2022) 26

Solutions 27

Ex 1 (from UKMT 2021) 28

Ex 2 (from OMB 2022) 28

Ex 3 (from UKMT 2019) 29

Ex 4 (from POFM 2015) 29

Ex 5 (from UKMT 2021) 30

Ex 6 (from AMC 2023) 30

Ex 7 (from POFM 2016) 31

Ex 8 (from PMO 2017) 31

Ex 9 (from POFM 2019) 32

Ex 10 (from UKMT 2021) 33

Ex 11 (from SMO 2021) 34

Ex 12 (from PMO 2009) 34

Ex 13 (from UKMT 2022) 35

Ex 14 (from POFM 2015) 35

Ex 15 (from UKMT 2021) 36

Ex 16 (from SMO 2022) 38

Ex 17 (from OMB 2020) 38

Ex 18 (from PMO 2016) 39

Ex 19 (from CMO 2019) 39

Ex 20 (from UKMT 2021) 40

Ex 21 (from SMO 2020) 41

Ex 22 (from AMC 2023) 42

Ex 23 (from POFM 2023) 43

Ex 24 (from UKMT 2021) 43

Ex 25 (from UKMT 2021) 44

Ex 26 (from SMO 2022) 45

Ex 27 (from PMO 2018) 46

Ex 28 (from OMS 2023) 47

Ex 29 (from AIME 2021) 48

Ex 30 (from POFM 2020) 49

Ex 31 (from Kangaroo 2023) 49

Ex 32 (from CJMO 2019) 49

Ex 33 (from AMC 2023) 50

Ex 34 (from SMO 2022) 50

Ex 35 (from UKMT 2021) 51

Ex 36 (from AMC 2022) 52

Ex 37 (from POFM 2014) 53

Ex 38 (from PMO 2017) 53

Ex 39 (from POFM 2022) 54

Ex 40 (from AIME2020) 55

Ex 41 (from SMO 2022) 57

Ex 42 (from Kangaroo 2015) 57

Ex 43 (from UKJMO 2015) 58

Ex 44 (from POFM 2015) 58

Ex 45 (from Maclaurin 2017) 59

Ex 46 (from AIME 2021) 61

Ex 47 (from RMC 2017) 63

Ex 48 (from PMO 2022) 64

Ex 49 (from CMO 2023) 65
Ex 50 (from AIME 2022) 67

Introduction

This book gathers 50 introductory-level exercises posed during different competitions (UKMT, SMO, AMC,...). All exercises are corrected in detail to assimilate the essential and useful notions to know.

This book naturally addresses middle/high school students, motivated by problem solving outside the school curriculum and/or looking to participate in various competitions to improve their college application. It will be particularly useful for those who are looking for training to discover the junior mathematical Olympiads or any other mathematical competition.

Here is the meaning of the acronyms used in the book:

UKMT: United Kingdom Mathematics Trust

SMO: Singapore Mathematical Olympiad

AMC: American Mathematics Competitions

OMB: Olympiade Mathématique Belge (Belgian Mathematical Olympiad)

POFM: Préparation Olympique Française de Mathématiques (French Mathematical Olympiad Preparation)

AIME: American Invitational Mathematics Examination

PMO: Philippine Mathematical Olympiad

CMO: Caucasian Mathematical Olympiad

OMS: Olympiade Mathematique Suisse (Swiss Mathematical Olympiad)

RMC: Romanian Mathematical Competitions

Here are some of the benefits of participating in junior Olympiads or any math contests:

- Develop a taste for mathematics: Junior Olympiads can help students develop a love of mathematics by exposing them to challenging problems that require creativity and problem-solving skills.

- Learn new mathematical concepts: Junior Olympiads often require students to learn new mathematical concepts that they may not have encountered in school. This can help students expand their mathematical knowledge and become more well-rounded mathematicians.

- Improve problem-solving skills: Junior Olympiads require students to develop strong problem-solving skills. This can be a valuable skill for students to have in any field, as it can help them think critically and creatively to solve problems.

- Gain confidence: participating in junior Olympiads can help students gain confidence in their mathematical abilities. This can be a valuable asset for students, as it can help them succeed in school and in other areas of their lives.

Here are some tips for succeeding in the Olympiads:

- Do a lot of exercises. The more exercises you do, the more comfortable you will be with the types of questions you will be asked.

- Learn the basics. Make sure you understand the basic concepts of mathematics, such as algebra, geometry, and trigonometry.

- Develop a problem-solving method. Develop a method that you can follow to solve mathematical problems. This method should include steps such as carefully reading the problem, identifying the knowledge and concepts needed to solve the problem, creating a plan to solve the problem, and solving the problem.

- Be patient and persistent. The Olympiads are difficult tests, but if you are patient and persistent, you can succeed.

All the best of luck in your preparations for the Olympiads!

Problems

Ex 1 (from UKMT 2021)
Evaluate:
$$\left(1+\frac{1}{1^2}\right)\left(2+\frac{1}{2^2}\right)\left(3+\frac{1}{3^2}\right)$$

Ex 2 (from OMB 2022)
What is the largest prime number that divides $10^{22} + 10^{23} + 10^{24}$?

Ex 3 (from UKMT 2019)
How many fractions between $\frac{1}{6}$ and $\frac{1}{3}$ inclusive can be written with a denominator of 15?

Ex 4 (from POFM 2015)
Let $a_n = \dfrac{1}{2n-1} - \dfrac{1}{2n+1}$

So, $a_1 = 1 - \dfrac{1}{3}$ and $a_2 = \dfrac{1}{3} - \dfrac{1}{5}$

If $S = a_1 + a_2 + a_3 + \ldots + a_{100}$

Evaluate $201 \times S$

Ex 5 (from UKMT 2021)

To travel the 140 kilometers between Glasgow and Dundee, John travels half an hour by bus and two hours by train. The train travels 20 km/h faster than the bus. The bus and the train both travel at constant speeds. What is the speed of the bus?

Ex 6 (from AMC 2023)

What is the value of

$$3 + \cfrac{1}{3 + \cfrac{1}{3 + \cfrac{1}{3 + \frac{1}{3}}}}$$

1. $\frac{31}{10}$
2. $\frac{49}{15}$
3. $\frac{33}{10}$
4. $\frac{109}{33}$
5. $\frac{15}{4}$

Ex 7 (from POFM 2016)

What is the value of a such as

$$\frac{1}{\sqrt{a+7}} + \frac{1}{7} = \frac{19}{84}$$

Ex 8 (from PMO 2017)
Find x if
$$\frac{79}{125}\left(\frac{79+x}{125+x}\right) = 1$$

1. 0
2. −46
3. −200
4. −204

Ex 9 (from POFM 2019)
On my board I wrote three numbers. By adding them two by two, I get the three sums: 69, 72, and 81. What was the greatest number written on the board?

Ex 10 (from UKMT 2021)
In my desk, the number of pencils and pens was in the ratio 4 : 5. I took out a pen and replaced it with a pencil and now the ratio is 7 : 8. What is the total number of pencils and pens in my desk?

Ex 11 (from SMO 2021)
Let $x = 2^{300}, y = 3^{200}, z = 6^{100}$.

Which of the following is true?

1. $x > y > z$
2. $x > z > y$
3. $y > z > x$
4. $y > x > z$
5. $z > x > y$

Ex 12 (from PMO 2009)
Simplify:

$$1 - \frac{1}{3} + \frac{1}{5} - \frac{1}{7} + \frac{1}{11}$$

Ex 13 (from UKMT 2022)
Two positive numbers a and b, with $a > b$, are such that twice their sum is equal to three times their difference. What is the ratio $a : b$?

Ex 14 (from POFM 2015)
Evaluate

$$\left(\frac{1+3^2}{\sqrt{21+\sqrt{16}}}\right)^5$$

Ex 15 (from UKMT 2021)
Amy, Bruce, Chris, Donna and Eve had a race. When asked in which order they finished, they all answered with a true and a false statement as follows:

Amy: Bruce came second and I finished in third place.

Bruce: I finished second and Eve was fourth.

Chris: I won and Donna came second.

Donna: I was third and Chris came last.

Eve: I came fourth and Amy won.

In which order did the participants finish?

Ex 16 (from SMO 2022)

Let a and b be real numbers satisfying $a < 0 < b$. Wich of the following is not true?

1. $a^2 b > 0$
2. $ab^2 < 0$
3. $\frac{a}{b} > 0$
4. $b - a > 0$
5. $|b - a| > 0$

Ex 17 (from OMB 2020)

If n is a nonzero natural number, $n!$ is an abbreviation for:

$$n(n-1)(n-2) \ldots \times 2 \times 1.$$

For example, $5! = 5 \times 4 \times 3 \times 2 \times 1 = 120$.

What is the smallest non-zero natural number whose product by $12!$ is a perfect square?

Ex 18 (from PMO 2016)

If

$$27^3 + 27^3 + 27^3 = 27^x$$

What is the value of x?

1. $\frac{10}{3}$
2. 4
3. 9
4. 12

Ex 19 (from CMO 2019)

In the kindergarten there is a big box with balls of three colors: red, blue and green, 100 balls in total. Once Pasha took out of the box 30 red, 10 blue, and 20 green balls and played with them. Then he lost five balls and returned the others back into the box. The next day, Sasha took out of the box 8 red, 18 blue, and 48 green balls. Is it possible to determine the color of at least one lost ball?

Ex 20 (from UKMT 2021)

The solution to each clue of this crossnumber is a two-digit number, that does not begin with a zero.

1	2
3	

Across

1. A prime

3. A square

Down

1. A square

2. A square

Find all the different ways in which the crossnumber can be completed correctly.

Ex 21 (from SMO 2020)

Wich of the 5 numbers has the largest value?

$2^{30}, 8^{19}, 4^{14}, 6^{12}, 9^{10}$

1. 2^{30}
2. 8^{19}
3. 4^{14}
4. 6^{12}
5. 9^{10}

Ex 22 (from AMC 2023)

The sum of three numbers is 96. The first number is 6 times the third number, and the third number is 40 less than the second number. What is the absolute value of the difference between the first and the second number?

Ex 23 (from POFM 2023)

Evaluate

$$\frac{4^8}{8^4}$$

Ex 24 (from UKMT 2021)
The product of five different integers is 12. What is the largest of the integers?

Ex 25 (from UKMT 2021)
The letters A, B and C stand for different, non-zero digits. Find all the possible solutions to the word-sum shown.

$$\begin{array}{r} ABC \\ + \;BCA \\ + \;CAB \\ \hline ABBC \end{array}$$

Ex 26 (from SMO 2022)
Evaluate

$$\sqrt{219 \times 220 \times 221 \times 222 + 1}$$

Ex 27 (from PMO 2018)
If 30% of p is q, and 20% of q is 12, what is 50% of p + q?

Ex 28 (from OMS 2023)
What is the smallest value that the following expression can take, if x is integer and $x \geq 42$?

$$\frac{2023}{1+\frac{1}{x}} + \frac{2023}{1+x}$$

Ex 29 (from AIME 2021)

Zou and Chou are practicing their 100meter sprints by running 6 races against each other. Zou wins the first race, and after that, the probability that one of them wins a race is 2/3 if they won the previous race but only 1/3 if they lost the previous race. The probability that Zou will win exactly 5 of the 6 races is m/n, where m and n are relatively prime positive integers. Find m+n.

Ex 30 (from POFM 2020)

Alexie and Baptiste each own a building. Each floor of Alexie's building has 3 bathrooms and 2 bedrooms. Baptiste has 4 bathrooms and 3 bedrooms per floor. There are in total (i.e. in the two buildings) 25 bathrooms and 18 bedrooms. Find the number of floors of Alexie's and Baptiste's building

Ex 31 (from Kangaroo 2022)

$6 \square 9 \square 12 \square 15 \square 18 \square 21 = 45$

In the equation there are five empty squares. Sanja wants to fill four of them with plus signs and one with a minus sign so that the equation is correct. Where should she place the minus sign?

- A. Between 6 and 9
- B. Between 9 and 12
- C. Between 12 and 15
- D. Between 15 and 18
- E. Between 18 and 21

Ex 32 (from CJMO 2019)

A function f is called injective if when f(n) = f(m), then n = m. Suppose that f is injective and

$$\frac{1}{f(n)} + \frac{1}{f(m)} = \frac{4}{f(n) + f(m)}$$

Prove m = n.

Ex 33 (from AMC 2023)

Two integers are inserted into the list 3,3,8,11,28 to double its range. The mode and median remain unchanged. What is the maximum possible sum of the two additional numbers?

A. 56

B. 57

C. 58

D. 60

E. 61

Ex 34 (from SMO 2022)

If a and b are distinct solutions to the equation

$$x^2 + 10x + 20 = 0$$

What is the value of $a^4 + b^4$?

Ex 35 (from UKMT 2021)

The pupils in my class work very quickly. Jasleen answers four questions every 30 seconds and Ella answers five questions every 40 seconds. Last week, Jasleen took exactly 1 hour to answer a large set of questions. How many minutes more than Jasleen did Ella take to answer the same set of questions?

A. 2
B. $2\frac{1}{2}$
C. $3\frac{1}{2}$
D. 4
E. $4\frac{1}{2}$

Ex 36 (from AMC 2022)

The sum of three numbers is 96. The first number is 6 times the third number, and the third number is 40 less than the second number. What is the absolute value of the difference between the first and second numbers?

Ex 37 (from POFM 2014)

a) Is 2016 divisible by 81? Is 20162016 divisible by 81?

b) Show that the integer N = 2016 · · · 2016 ("2016" being written 2016 times) is divisible by 81.

Ex 38 (from PMO 2023)

Suppose $a_1 < a_2 < \cdots < a_{25}$ are positive integers such that the average of $a_1, a_2, \ldots, a_{24}$ is one-half the average of $a_1, a_2, \ldots, a_{25}$. What is the minimum possible value of a25?

5. 26
6. 275
7. 299
8. 325

Ex 39 (from POFM 2022)

Theo received his grades for the term, which are all integers between 1 and 5 (1 and 5 included). He notes that the average of his marks is less than or equal to 3. Thus, so that he is not deprived of dessert for a week, he plans to replace on his transcript all his notes equal to 1 by as many notes equal to 3. Show that after this transformation, the average scores remains less than or equal to 4.

Ex 40 (from AIME2020)

A positive integer N has base-eleven representation $\underline{a\,b\,c}$ and base-eight representation $\underline{1\,b\,c\,a}$, where a, b and c represent (not necessarily distinct) digits. Find the least such N expressed in base ten.

Ex 41 (from SMO 2022)

Consider the following product of two mixed fractions

$$m\frac{6}{7} \times n\frac{1}{3} = 23$$

Where m and n are positive integers. What is the value of $m + n$?

Ex 42 (from Kangaroo 2015)
What is the remainder when
$$2^2 \times 3^3 \times 5^5 \times 7^7$$
is divided by 8?

- A. 2
- B. 3
- C. 4
- D. 5
- E. 7

Ex 43 (from UKJMO 2015)
My fruit basket contains apples and oranges. The ratio of apples to oranges in the basket is 3 : 8. When I remove one apple the ratio changes to 1 : 3.

How many oranges are in the basket?

Ex 44 (from POFM 2017)
Determine a, a number a such that $85a - 2630$ is a prime number.

Ex 45 (from MacLaurin 2017)
Solve the pair of simultaneous equations
$$\begin{cases} (a+b)(a^2 - b^2) = 4 \\ (a-b)(a^2 + b^2) = \dfrac{5}{2} \end{cases}$$

Ex 46 (from AIME 2021)
Find the number of ways 66 identical coins can be separated into three nonempty piles so that there are fewer coins in the first pile than in the second pile and fewer coins in the second pile than in the third pile.

Ex 47 (from RMC 2017)
Consider the positive real numbers x, y, z, with the property that:
$$xy = \frac{z - x + 1}{y} = \frac{z + 1}{2}$$
Prove that x is the average of y and z.

Ex 48 (from PMO 2022)
How many five-digit numbers containing each of the digits 0, 1, 2, 3, 4, 5 taken five at a time are divisible by 25?

5. 2
6. 32
7. 42
8. 52

Ex 49 (from CMO 2023)
Aroon's PIN has four digits. When the first digit (reading from the left) is moved to the end of the PIN, the resulting integer is 6 less than 3 times Aroon's PIN. What could Aroon's PIN be?

Ex 50 (from AIME 2022)

Adults made up $\frac{5}{12}$ of the crowd of people at a concert. After a bus carrying 50 more people arrived, adults made up $\frac{11}{25}$ of the people at the concert. Find the minimum number of adults who could have been at the concert after the bus arrived.

Solutions

Ex 1 (from UKMT 2021)

$$\left(1+\frac{1}{1^2}\right)\left(2+\frac{1}{2^2}\right)\left(3+\frac{1}{3^2}\right) =$$

$$\left(1+\frac{1}{1}\right)\left(2+\frac{1}{4}\right)\left(3+\frac{1}{9}\right) =$$

$$2 \times \frac{9}{4} \times \frac{28}{9} = 14$$

Ex 2 (from OMB 2022)

$$10^{22} + 10^{23} + 10^{24} =$$

$$10^{22}(1 + 10 + 100) =$$

$$10^{22} \times 111 =$$

Notice

$$111 = 3 \times 37$$

And

$$10^{22} = 2^{22} \times 5^{22}$$

So,

$$10^{22} + 10^{23} + 10^{24} = 2^{22} \times 5^{22} \times 3 \times 37$$

The largest divisor prime number of $10^{22} + 10^{23} + 10^{24}$ is 37

Ex 3 (from UKMT 2019)
By reducing these fractions to the same denominator, we get:

$$\frac{1}{6} = \frac{5}{30}$$

$$\frac{1}{3} = \frac{10}{30}$$

We are therefore looking for fractions that are written with a denominator of 15, so of the form $\frac{n}{30}$, with n an even number.

To have such a fraction between $\frac{5}{30}$ et $\frac{10}{30}$, n can only be 6, 8 or 10.

So :

$$\frac{6}{30} = \frac{3}{15}$$

$$\frac{8}{30} = \frac{4}{15}$$

$$\frac{10}{30} = \frac{5}{15}$$

Ex 4 (from POFM 2015)

$$S = 1 - \frac{1}{3} + \frac{1}{3} - \frac{1}{5} + \frac{1}{5} - \frac{1}{3} - \frac{1}{7} + \cdots + \frac{1}{199} - \frac{1}{201}$$

$$S = 1 - \frac{1}{201}$$

Finally, $201S = 201 - 1 = 200$

Ex 5 (from UKMT 2021)

Let v be the speed of the bus (km/h). The speed of the train is $(v + 20)$ km/h. Therefore,

$$\frac{1}{2} \times v + 2(v + 20) = 140$$

$$\frac{5v}{2} + 40 = 140$$

$$v = \frac{2 \times 100}{5} = 40$$

Ex 6 (from AMC 2023)

We start with the operation located at the bottom:

$$3 + \frac{1}{3} = \frac{10}{3}$$

Then, plug in the whole operation:

$$3 + \cfrac{1}{3 + \cfrac{1}{3 + \cfrac{1}{3 + \frac{1}{3}}}} = 3 + \cfrac{1}{3 + \cfrac{1}{3 + \frac{1}{10/3}}} = 3 + \cfrac{1}{3 + \cfrac{1}{3 + \frac{3}{10}}}$$

$$3 + \cfrac{1}{3 + \cfrac{3}{3 + \frac{3}{10}}} = 3 + \cfrac{1}{3 + \cfrac{1}{\frac{30+3}{10}}} = 3 + \cfrac{1}{3 + \cfrac{1}{\frac{33}{10}}} =$$

$$3 + \frac{10}{33} = \frac{99 + 10}{33} = \frac{109}{33}$$

Ex 7 (from POFM 2016)

$$\frac{1}{\sqrt{a+7}} = \frac{19}{84} - \frac{1}{7} = \frac{19-12}{84} = \frac{7}{84} = \frac{1}{12}$$

Performing a cross product, we get:

$$\sqrt{a+7} = 12$$

By squaring each term:

$$a + 7 = 144$$

Hence,

$$a = 144 - 7 = 137$$

Ex 8 (from PMO 2017)

Any non-zero number multiplied by its inverse equals 1, so

①$\begin{cases} 79 + x = 125 \\ 125 + x = 79 \end{cases}$ and ② $\begin{cases} 79 + x = -125 \\ 125 + x = -79 \end{cases}$

Solving these two systems gives

For ①

$x = 125 - 79 = 46$
$x = 79 - 125 = -46$ (which is impossible)

For ②

$$\begin{cases} x = -125 - 79 = -206 \\ x = -79 - 125 = -206 \end{cases}$$

Ex 9 (from POFM 2019)

Let a, b, and c be the three numbers noted in the table

$$\begin{cases} a + b = 69 & ① \\ b + c = 72 & ② \\ a + c = 81 & ③ \end{cases}$$

$$\begin{cases} a + b = 69 & ① \\ a + b - b - c = 72 & ① - ② \\ a + c = 81 & ③ \end{cases}$$

$$\begin{cases} a + b = 69 & ① \\ a - c = -3 & ① - ② \\ a + c = 81 & ③ \end{cases}$$

$$\begin{cases} a + b = 69 & ① \\ a - c = 72 & ① - ② \\ a - c + a + c = 78 & ① - ② + ③ \end{cases}$$

$$\begin{cases} a + b = 69 & ① \\ a - c = 72 & ① - ② \\ 2a = 78 & ① - ② + ③ \end{cases}$$

We deduce that a= 39

By replacing we get :

$$39 + b = 69 \quad so, b = 30$$
$$30 + c = 72 \quad so, c = 42$$

Ex 10 (from UKMT 2021)

At the beginning, the ratio of the number of pencils to the number of pens was 4 : 5.

Let :

- 4p, be the number of pencils
- 5p, be the number of pens

When you replaced a pen with a pencil, the total number (pens + pencils) remains the same but the ratio will be 7 : 8.

$$\frac{4p + 1}{5p - 1} = \frac{7}{8}$$

Performing a cross product, we get:

$$7(5p - 1) = 8(4p + 1)$$
$$35p - 7 = 32p + 8$$
$$3p = 15$$
$$p = 5$$

The total number of pens and pencils on my desk is:

$$4 \times 5 + 5 \times 5 = 45$$

Note that the total number (pens and pencils) on the desk does not vary (a pen is replaced by a pencil).

Ex 11 (from SMO 2021)

We need to write these three numbers with the same power using the power formula:

$$a^{m \times n} = (a^m)^n$$

So, we have:

$$x = 2^{300} = 2^{3 \times 100} = (2^3)^{100} = 8^{100}$$

$$y = 3^{200} = 3^{2 \times 100} = (3^2)^{100} = 9^{100}$$

$$z = 6^{100}$$

Therefore, the answer is 4.(, $y > x > z$)

Ex 12 (from PMO 2009)

$$1 - \frac{1}{3} + \frac{1}{5} - \frac{1}{7} + \frac{1}{11} =$$

All denominators are prime numbers:

$$\frac{3 \times 5 \times 7 \times 11 - 5 \times 7 \times 11 + 3 \times 7 \times 11 - 3 \times 5 \times 11 + 3 \times 5 \times 7}{3 \times 5 \times 7 \times 11} =$$

By factoring 3 terms of the numerator by 7×11 we get:

$$\frac{7 \times 11(15 - 5 + 3) - 15 \times 11 + 15 \times 7}{3 \times 5 \times 7 \times 11} =$$

By factoring 2 terms of the numerator by 15:

$$\frac{7 \times 11 \times 13 + 15(-11 + 7)}{3 \times 5 \times 7 \times 11} =$$

$$\frac{7 \times 11 \times 13 + 15 \times (-4)}{3 \times 5 \times 7 \times 11} = \frac{1001 - 60}{1155} = \frac{941}{1155}$$

Ex 13 (from UKMT 2022)

" a and b, with $a > b$, are such that twice their sum is equal to three times their difference" can be translated in mathematics by:

$$2(a + b) = 3(a - b)$$

By expanding:

$$2a + 2b = 3a - 3b$$

We end up with

$$5b = a$$

the ratio $a : b$ is $5 : 1$

Ex 14 (from POFM 2015)

$$\left(\frac{1 + 3^2}{\sqrt{21 + \sqrt{16}}}\right)^5 =$$

$$\left(\frac{1 + 9}{\sqrt{21 + 4}}\right) = ^5$$

$$\left(\frac{10}{\sqrt{25}}\right)^5 =$$

$$\left(\frac{10}{5}\right)^5 =$$

$$2^5 = 32$$

Ex 15 (from UKMT 2021)

We keep in mind that each person made a true statement and a false statement.

Assume that Amy's first answer (Bruce came in second) is true.

If Amy's first answer is true then:

- Bruce's first answer is true (I finished second)

- Bruce's second answer is wrong (Eve was fourth)

If Eve was not fourth then:

- Eve's first answer is wrong (I finished fourth)

- Eve's second answer is true (Amy won)

So if Bruce came second then Amy won, which means that both of Chris' answers (I won and Donna came second) are wrong, which is impossible.

Which means Amy's first answer (Bruce came in second) is wrong and Amy's second (I came in third) is right.

So Amy came in third

If Amy is third then:

- Eve's first answer is true (I finished fourth)

- Eve's second answer is wrong (Amy won)

So, Eve came fourth

Also, if Amy is third then:

- Dona's first answer is wrong (I was third)

- Dona's second answer is true (Amy won)

If Eve is fourth then:

- Eve's first answer is wrong (I finished fourth)

- Eve's second answer is true (Chris came last)

Chris is last

If Chris is last then:

- Chris' first answer is wrong (I won)

- Chris's second answer is true (Donna came second)

Dona is therefore second

The order is:

Bruce, Dona, Amy, Eve, and Chris

Ex 16 (from SMO 2022)
If $a < 0 < b$ then:

1. $a^2 > 0\ then\ a^2 b > 0$ True (multiplication of two positive numbers)
2. $b^2 > 0\ then\ a^2 b < 0$ True (multiplication of one positive number by one negative number)
3.
4. $\frac{a}{b} < 0$ False (division of one positive number by one negative number)
5.
6. $b > a\ then\ b - a > 0$ True
7. $|b - a| > 0$ True

Ex 17 (from OMB 2020)
By decomposing 12! into prime factors, we get:

$$12! = 12 \times 11 \times 10 \times 9 \times 8 \times 7 \times 6 \times 5 \times 4 \times 3 \times 2 \times 1$$

$$12! = 2^{10} \times 3^5 \times 5^2 \times 7 \times 11.$$

2^{10} and 5^2 are already perfect squares, so we just need to multiply this number by:

$$3 \times 7 \times 11 = 231$$

to get the smallest non-zero perfect square possible!

Ex 18 (from PMO 2016)

$$27^3 + 27^3 + 27^3 = 27^x$$

$$27^3 \times 3 = 27^x$$

$$(3^3)^3 \times 3 = (3^3)^x$$

Using the power rule, we can simplify the expression as follows:

$$(a^n)^m = a^{n \times m}$$

$$3^9 \times 3 = 3^{3x}$$

Using another power rule, we can simplify the expression as follows:

$$a^m \times a^n = a^{m+n}$$

We get

$$3^{10} = 3^{3x}$$

$$3x = 10$$

Hence,

$$x = \frac{10}{3}$$

Ex 19 (from CMO 2019)

Yes, we can conclude that at least one lost ball is red and here is why:

Sasha took 18 blue and 48 green balls:

$$18 + 48 = 66$$

The minimum non-red balls in the box was 66.

Assume that no red balls were lost by Pasha:

Then, 30 red balls would be in the box:

$$66 + 30 = 96$$

that means 96 balls would be in the box. The statement of the problem gives us 95, so our assumption is wrong (no red balls were lost by Pasha).

So, Pasha lost at least one red ball

Ex 20 (from UKMT 2021)

First, we know that the two-digit perfect squares are 16, 25, 36, 49, 64, and 81.

"2" Down and "3" Across are squares with the same last digit in common, so 6.

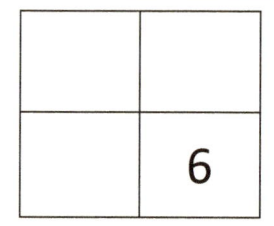

	6

"3" Across ends in 6 and is a perfect square, so 16 or 36. "1" Down is a perfect square ending in 1 or 3, so the "3" Across is 16 and «1" Down is 81.

8	
1	6

"2" Down is a square ending in 6 (so 16 or 36), "1" Across being prime (81 is not prime), so it is equal to 83.

8	3
1	6

Ex 21 (from SMO 2020)
Using the following power rule:
$$(a^n)^m = a^{n \times m}$$

Let's compare the first 2 numbers:
$$2^{30} = 2^{3 \times 10} = (2^3)^{10} = 8^{10}$$
$$8^{10} > 8^{19}$$

The first one with the third one:
$$4^{14} = (2^2)^{14} = 2^{28}$$
$$2^{30} > 2^{28}$$

The fourth one with the fifth one:
$$9^{10} = (3^2)^{10} = 3^{20} = 3^{12} \times 3^8$$
$$6^{12} = (3 \times 2)^{12} = 3^{12} \times 2^{12}$$
$$3^8 = (3^2)^4 = 9^4$$
$$2^{12} = (2^3)^4 = 8^4$$

So, $9^4 > 8^4$

Hence :
$$9^{10} > 6^{12}$$

And finally, The first one with the fifth one:

41

$$2^{30} = 2^{3 \times 10} = (2^3)^{10} = 8^{10}$$

We conclude that :

$$9^{10} > 2^{30}$$

9^{10} has the largest value

Ex 22 (from AMC 2023)

Let $a, b \text{ and } c$ be these three numbers with the sum equals to 96.

$$a + b + c = 96 \; (①)$$

The first number is 6 times the third number:

$$a = 6c$$

The third number is 40 less than the second number:

$$c = b - 40$$
$$b = c + 40$$

Expressing equation ① function of c:

$$6c + c + 40 + c = 96$$
$$8c = 56$$
$$c = 7$$

Plugging the value of c to get the value of b:

$$a = 6c = 42$$
$$b = c + 40 = 47$$

Hence, :

$$|a - b| = |42 - 47| = |-5| = 5$$

Ex 23 (from POFM 2023)

$$\frac{4^8}{8^4} = \frac{(2^2)^8}{(2^3)^4}$$

Using the following power rule

$$(a^n)^m = a^{n \times m}$$

$$\frac{4^8}{8^4} = \frac{(2^2)^8}{(2^3)^4} = \frac{2^{2 \times 8}}{2^{3 \times 4}} = \frac{2^{16}}{2^{12}}$$

Using the following power rule

$$\frac{a^m}{a^n} = a^{m-n}$$

$$\frac{4^8}{8^4} = \frac{(2^2)^8}{(2^3)^4} = \frac{2^{2 \times 8}}{2^{3 \times 4}} = \frac{2^{16}}{2^{12}} = 2^{16-12} = 2^4 = 16$$

Ex 24 (from UKMT 2021)

If you consider the smallest first five positive integers the product is:

$$1 \times 2 \times 3 \times 4 \times 5 = 120$$

We need to remind that an integer can be positive or negative. But we need an even number of negative integers to end up with a positive result. The smallest positive product of four different integers is:

$$(-2) \times (-1) \times 1 \times 2 = 4$$

We can deduce that the only five different integers whose product is 12 are −2, −1, 1, 2, and 3.

$$-2 \times -1 \times 1 \times 2 \times 3 = 12$$

Ex 25 (from UKMT 2021)

We add three three-digit numbers, so the result is less than 3000, which gives $A = 1$ or $A = 2$.

We look at the columns one by one, starting with the units, then the tens, and finally the hundreds.

With the unit's column, we can deduce that, since A and B are non-zero and $A + B < 20$, $C + A + B = 10 + C$.

So $A + B = 10$

With the ten's column, we start by carrying over the remainder (equal to 1).

$$1 + B + C + A = 10 + B.$$

So $C + A = 9$

With the hundred's column, we start by carrying over the remainder (equal to 1). We notice that there is a remainder of 1 in the thousand's column, so $A = 1$.

$$A + B = 10$$
$$B = 9$$
$$C + A = 9$$
$$C = 8$$

In the end, we get:

$$A = 1$$
$$B = 9$$
$$C = 8$$

So the only possible solution is:

$$198$$
$$+\,981$$
$$+\,819$$
$$\overline{1998}$$

Ex 26 (from SMO 2022)

Let x be :

$$219 = x$$

By expressing the other terms function of x:

$$220 = x + 1$$
$$221 = x + 2$$
$$222 = x + 3$$

$$\sqrt{219 \times 220 \times 221 \times 222 + 1} =$$
$$\sqrt{x(x+1)(x+2)(x+3) + 1} =$$

By multiplying and expanding the first and the fourth term:

$$x(x+3) = (x^2 + 3x)$$

By multiplying and expanding the second and the third term:

$$(x+1)(x+2) = (x^2 + 3x + 2)$$

$$\sqrt{x(x+1)(x+2)(x+3)+1} =$$

$$\sqrt{(x^2+3x)(x^2+3x+2)+1} =$$

Let X be: $X = x^2 + 3x$

$$\sqrt{X(X+2)+1} = \sqrt{X+2X+1} = \sqrt{(X+1)^2} =$$

$$X + 1 = x^2 + 3x + 1 = 219^2 + 3 \times 219 + 1 = 48619$$

Ex 27 (from PMO 2018)

« 20% of q is 12 » can be expressed as:

$$0{,}2q = 12$$

So,

$$q = 60$$

« 30% of p is q » can be expressed as:

$$q = 0{,}3p$$

Plugging the value of q we get:

$$60 = 0{,}3\,p$$

$$p = \frac{60}{0{,}3} = 200$$

« 50% of p + q » can be expressed as:

$$0{,}5(p+q)$$

Hence,

$$0{,}5(p+q) = 0{,}5(200+60) = 130$$

Ex 28 (from OMS 2023)

$$\frac{2023}{1+\frac{1}{x}} + \frac{2023}{1+x}$$

Reducing to the same denominator:

$$\frac{2023}{1+\frac{1}{x}} + \frac{2023}{1+x} =$$

$$\frac{2023(1+x) + 2023\left(1+\frac{1}{x}\right)}{\left(1+\frac{1}{x}\right)(1+x)} =$$

Factor by 2023:

$$\frac{2023\left[(1+x)\left(1+\frac{1}{x}\right)\right]}{\left(1+\frac{1}{x}\right)(1+x)} =$$

$$2023$$

The result is equal to 2023 regardless of the value of x, so the smallest value the expression can take is 2023.

Ex 29 (from AIME 2021)

For the next five races, Zou wins four and loses one.

Let W be a won race, and L a lost race for Zou.

The possible outcomes (for the remaining 5 races) are:

WWWWL, WWWLW, WWLWW, WLWWW, LWWWW

The probability of the first outcome is: $\left(\frac{2}{3}\right)^4 \times \left(\frac{1}{3}\right) = \frac{16}{3^5}$

The probability of the second outcome is: $\left(\frac{2}{3}\right)^3 \times \left(\frac{1}{3}\right) \times \left(\frac{1}{3}\right) = \frac{8}{3^5}$

The probability of the third outcome is: $\left(\frac{2}{3}\right)^2 \times \left(\frac{1}{3}\right) \times \left(\frac{1}{3}\right) \times \left(\frac{2}{3}\right) = \frac{8}{3^5}$

The probability of the fourth outcome is: $\left(\frac{2}{3}\right) \times \left(\frac{1}{3}\right) \times \left(\frac{1}{3}\right) \times \left(\frac{2}{3}\right)^2 = \frac{8}{3^5}$

The probability of the fifth: $\left(\frac{1}{3}\right) \times \left(\frac{1}{3}\right) \times \left(\frac{2}{3}\right)^3 = \frac{8}{3^5}$

By adding these five probabilities, we get the total probability which is:

$$\frac{16 + 8 \times 4}{3^5} = \frac{48}{243} = \frac{16}{81}$$

So, $m + n = 16 + 81 = 97$

Ex 30 (from POFM 2020)

Let x be the number of floors in Alexie's building and y the number of floors in Baptiste's building. We can express the statement as a system of equations:

$$\begin{cases} 3x + 4y = 25 & \text{①} \\ 2x + 3y = 18 & \text{②} \end{cases}$$

2① − 3②: :

$$6x + 8y - 6x - 9y = 50 - 54$$

$$y = 4$$

By plugging the value of y in ①, we get:

$$3x + 4(4) = 25$$

$$x = 3$$

Alexie's building has 3 floors and Baptiste's 4.

Ex 31 (from Kangaroo 2023)

$6 + 9 + 12 + 15 + 18 + 21 = 81$

And $81 - 45 = 36 = 2 \times 18$

Sanja needs to subtract rather than add 18.

So, the minus sign should be placed between 15 and 18 (answer D)

Ex 32 (from CJMO 2019)

$$\frac{f(m) + f(n)}{f(n)f(m)} = \frac{4}{f(n) + f(m)}$$

Clearing denominators gives

49

$$[f(n) + f(m)][f(n) + f(m)] = 4f(n)f(m)$$
$$f(n)^2 + 2f(n)f(m) + f(m)^2 = 4f(n)f(m)$$
$$f(n)^2 - 2f(n)f(m) + f(m)^2 = 0$$
$$(f(n) - f(m))^2 = 0$$

Thus, $f(n) = f(m)$, and since f is injective, so $m = n$.

Ex 33 (from AMC 2023)

Range: $28 - 3 = 25$

The double of the range is $2 \times 25 = 50$

The list is list3,3,8,11,28 , so the median is 8, this implies that one number is smaller than 8 the other one greater that 8(8 will modify the mode)

53 (because 53-3=50), is the first number of the new list and the second number is 7.

$$53 + 7 = 60$$

Answer D

Ex 34 (from SMO 2022)

If a and b are distinct solutions to the equation

$$x^2 + 10x + 20 = 0$$

Then

$$x^2 - (a+b)x + ab = 0$$

Thus,

$$\begin{cases} a + b = -10 \\ ab = 20 \end{cases}$$

$$(a + b)^2 = a^2 + b^2 + 2ab$$
$$a^2 + b^2 = (a + b)^2 - 2ab$$
$$a^2 + b^2 = (-10)^2 - 2(20)$$
$$a^2 + b^2 = 60$$

And

$$(a^2 + b^2)^2 = a^4 + b^4 + 2a^2b^2$$
$$a^4 + b^4 = (a^2 + b^2)^2 - 2(ab)^2$$
$$a^4 + b^4 = (60)^2 - 2(20)^2$$
$$a^4 + b^4 = 2800$$

Ex 35 (from UKMT 2021)

$$1 \text{ hour} = 60 \times 60 = 120 \times 30$$

Because, Jasleen answers four questions every 30 seconds, she answers $4 \times 120 = 480$ questions.

Ella answers five questions every 40 seconds

$$\frac{40}{5} = 8$$

Thus, 1 question every 8 seconds

Ella answers 480 questions in:

$$480 \times 8 = 3840 \text{ seconds}$$

Ella takes $3840 - 3600 = 240$ more seconds than Jasleen.

240 seconds = 4 minutes

Answer D

Ex 36 (from AMC 2022)
Let x be the third number.

Then $6x$ is the first number

And $x + 40$ is the second number

The sum is:

$$6x + (x + 40) + x = 96$$
$$8x + 40 = 96$$
$$x = 7$$

The absolute value of the difference between the first and second numbers is:

$$|(6 \times 7) - (7 + 40)| =$$
$$|42 - 47| =$$
$$|-5| = 5$$

Ex 37 (from POFM 2014)

b) Show that the integer N = 2016 · · · 2016 ("2016" being written 2016 times) is divisible by 81.

First of all, we recall that an integer is divisible by 9 if, and only if the sum of its numbers is. In particular, let us already notice that 2016 is divisible by 9.

Let us also recall the following result: let n, a, b be positive integers such that a divides n. If b divides n/a, then ab divides n. Conversely, if b does not divide n/a, then ab does not divide n.

a) A little calculation shows that 2016/9 = 224. The sum of the digits of 224 is not divisible by 9. We deduce that 2016 is not divisible by 9 × 9 = 81. Likewise, we see that 20162016/9 = 2240224. This number is not divisible by 9, so 20162016 is not divisible by 81.

b) We also see, by applying the division, that N/9 is equal to 0224 · · · 0224 ("0224" being written 2016 times).

We must show that this number is divisible by 9. However, the sum of the digits of this number is:

$$(2 + 2 + 4) \times 2016 = 8 \times (9 \times 224)$$

which is divisible by 9, therefore N/9 = 0224 · · · 0224 is indeed divisible by 9. We deduce that N is divisible by 9 · 9 = 81

Ex 38 (from PMO 2017)

"the average of $a_1, a_2, \ldots, a_{24}$ is one-half the average of $a_1, a_2, \ldots, a_{25}$" can be expressed as:

$$\frac{a_1 + a_2 + \cdots + a_{24}}{24} = \frac{1}{2} \times \frac{a_1 + a_2 + \cdots + a_{25}}{25}$$

$$a_1 + a_2 + \cdots + a_{24} = \frac{12}{25} \times (a_1 + a_2 + \cdots + a_{24} + a_{25})$$

$$\frac{12}{25} a_{25} = (a_1 + a_2 + \cdots + a_{24})\left(1 - \frac{12}{25}\right)$$

$$\frac{12}{25} a_{25} = \frac{13}{25}(a_1 + a_2 + \cdots + a_{24})$$

$$a_{25} = \frac{325}{300}(a_1 + a_2 + \cdots + a_{24})$$

Because, $a_1 < a_2 < \cdots < a_{24}$, the minimum value of $a_1 + a_2 + \cdots + a_{24}$ is $1 + 2 + 3 + \cdots + 24$ which is $\frac{24 \times 25}{2} = 25 \times 12 = 300$

Hence, the minimum value of a_{25}:

$$a_{25} = \frac{325}{300} \times 300 = 325$$

Answer D

Ex 39 (from POFM 2022)

Let a be the number of notes 1, b the number of notes 2, c the number of notes 3, d the number of notes 4, e the number of notes 5.

Let n the total number of notes.

$$n = a + b + c + d + e$$

The total is

$$\bar{x}_1 = \frac{a + 2b + 3c + 4d + 5e}{n}$$

When changing, notes 1 become 3,

The new average is

$$\bar{x}_2 = \frac{2b + 3(a+c) + 4d + 5e}{n}$$

$$\bar{x}_2 = \frac{a + 2b + 3c + 4d + 5e}{n} + \frac{2a}{n}$$

$$\bar{x}_2 = \bar{x}_1 + \frac{2a}{n}$$

Suppose that the mean after transformation is strictly greater than 4. Then the average has increased by strictly more than 1,

$$\frac{2a}{n} > 1$$

But,

$$\frac{a + 2b + 3c + 4d + 5e}{n} \leq \frac{3a + 5b + 5c + 5d + 5e}{n}$$

$$\frac{3a + 5b + 5c + 5d + 5e}{n} = \frac{3a - 5(n-a)}{n}$$

$$\frac{3a - 5(n-a)}{n} = 5 - \frac{2a}{n}$$

But, $\frac{2a}{n} > 1$, so, $5 - \frac{2a}{n} < 4$

Ex 40 (from AIME2020)

A positive integer N has base-eleven representation $\underline{a\ b\ c}$ and base-eight representation $\underline{1\ b\ c\ a}$, where a, b and c represent (not necessarily distinct) digits. Find the least such N expressed in base ten.

N in a base-eleven representation:

$$11^2 a + 11a + c = N$$
$$121a + 11a + c = N$$

N in a base-eight representation:

$$1 \times 8^3 + 8^2 b + 8c + a = N$$
$$512 + 64b + 8c + a = N$$

Note: The maximum digit in base eight is 7

So,

$$121a + 11a + c = 512 + 64b + 8c + a$$
$$120a = 512 + 53b + 7c$$
$$120a \geq 512$$

a is 5, 6, or 7

If $a = 5$

Then, $600 = 512 + 53b + 7c$

$$88 = 53b + 7c$$

Hence b=0 or 1:

if $b = 0$, then $c = \frac{88}{7}$ which is not an integer

if $b = 1$, then $c = \frac{88-53}{7} = 5$

Hence, 621 is the least such number.

Ex 41 (from SMO 2022)

$$\frac{7m+6}{7} \times \frac{3n+1}{3} = 23$$

$$(7m+6)(3n+1) = 23 \times 7 \times 3$$

m, n are positive integers so,

The minimum of $7m + 6$ is 13 and the minimum of $3n + 1$ is 4

So,

$$(7m+6)(3n+1) = 3 \times 23 \times 7$$

$$69 \times 7 = 3 \times 23 \times 7$$

$$\begin{cases} 7m + 6 = 69 \\ 3n + 1 = 7 \end{cases}$$

$$\begin{cases} m = 9 \\ n = 2 \end{cases}$$

Hence, $m + n = 11$

Ex 42 (from Kangaroo 2015)

$3^3, 5^5$ and 7^7 are 3 odd numbers, so, $3^3 \times 5^5 \times 7^7$ is an odd number and can be written: $2n + 1$.

$$2^2 \times 3^3 \times 5^5 \times 7^7 = 4(2n+1) = 8n + 4$$

Therefore when $2^2 \times 3^3 \times 5^5 \times 7^7$ is divided by 8, the quotient is n and the remainder is 4.

Answer C

Ex 43 (from UKJMO 2015)

Suppose there are initially x apples and y oranges in the basket.

Because $x : y = 3 : 8$, or

$8x = 3y$ ①

"When I remove one apple the ratio changes to 1 : 3" so, the ratio becomes $(x - 1) : y = 1 : 3$, or

$3(x - 1) = y$ ②

$$8x = 3y$$
$$8x = 3 \times 3(x - 1)$$
$$8x = 9x - 9$$
$$x = 9$$

Plugging the value of x in ②:

$$3(x - 1) = y$$
$$3(9 - 1) = y$$
$$y = 24$$

My fruit basket contains 24 oranges.

Ex 44 (from POFM 2015)

$$85a - 2630 = 5(17a - 526)$$

$5(17a - 526)$ is a multiple of 5 (not a prime number), if (and only if) $17a - 526 \neq 1$.

If $17a - 526 = 1 \rightarrow a = \dfrac{527}{17} = 31$

Ex 45 (from Maclaurin 2017)

$$a^2 - b^2 = (a+b)(a-b)$$

The first equation becomes:

$$(a+b)^2(a-b) = 4 \quad \text{①}$$

The second equation is:

$$(a-b)(a^2+b^2) = \frac{5}{2} \quad \text{②}$$

we can divide ① by ② to get

$$\frac{(a+b)^2(a-b)}{(a-b)(a^2+b^2)} = \frac{4}{\frac{5}{2}}$$

Hence:

$$\frac{(a+b)^2}{(a^2+b^2)} = \frac{8}{5}$$

$$5(a+b)^2 = 8(a^2+b^2)$$
$$5a^2 + 5b^2 + 10ab = 8a^2 + 8b^2$$
$$3a^2 - 10ab + 3b^2 = 0$$
$$(a-3b)(3a-b) = 0$$

First case: $a - 3b = 0$

We substitute $a = 3b$ into the first equation

$$4b \times 8b^2 = 4$$
$$32b^3 = 4$$

59

$$b^3 = \frac{1}{8}$$
$$b = \frac{1}{2}$$

We substitute $b = \frac{1}{2}$

$$a - \frac{3}{2} = 0$$
$$a = \frac{3}{2}$$

or $a - 3b = 0$ or $b = 3a$.

Second case: $3a - b = 0$

We substitute $b = 3a$ into the first equation

$$4a \times -8a^2 = 4$$
$$-32a^3 = 4$$
$$a^3 = -\frac{1}{8}$$
$$a = -\frac{1}{2}$$

We substitute $a = -\frac{1}{2}$

$$-\frac{3}{2} + b = 0$$
$$b = \frac{3}{2}$$

$$\left(\frac{3}{2}, \frac{1}{2}\right) \text{ or } \left(-\frac{3}{2}, -\frac{1}{2}\right)$$

Ex 46 (from AIME 2021)

Let a the coin(s) int the pile 1, b the coin sin the pile 2 and c, the coins in the pile 3.

$$0 < a < b < c$$

Let $k_1 \geq 1$ such as $\quad b = a + k_1$
Let $k_2 \geq 1$ such as $\quad c = b + k_2 = a + k_1 + k_2$

There are 66 identical coins, hence

$$a + b + c = 66$$
$$a + a + k_1 + a + k_1 + k_2 = 66$$
$$3a + 2k_1 + k_2 = 66$$

If $a = 1$:

$$2k_1 + k_2 = 63$$

$k_1 \geq 1$ and $k_2 \geq 1$, hence,

$$31 \geq k_1 \geq 1$$

If $k_1 = 1$, then $k_2 = 61$
If $k_1 = 2$, then $k_2 = 59$
If $k_1 = 3$, then $k_2 = 57$

...

For each value of k_1, we have a unique value of k_2, so, 31 solutions for $a = 1$

If $a = 2$:

$$2k_1 + k_2 = 60$$

$k_1 \geq 1$ and $k_2 \geq 1$, hence,
$$29 \geq k_1 \geq 1$$
If $k_1 = 1$, then $k_2 = 58$
If $k_1 = 2$, then $k_2 = 56$
If $k_1 = 3$, then $k_2 = 54$

For each value of k_1, we have a unique value of k_2, so, 29 solutions for $a = 2$

If $a = 3$:
$$2k_1 + k_2 = 57$$
$k_1 \geq 1$ and $k_2 \geq 1$, hence,
$$28 \geq k_1 \geq 1$$
If $k_1 = 1$, then $k_2 = 55$
If $k_1 = 2$, then $k_2 = 53$
If $k_1 = 3$, then $k_2 = 51$
...

For each value of k_1, we have a unique value of k_2, so, 28 solutions for $a = 3$

The number of solutions is all numbers (not multiples of 3) that are below or equal to 31

Number of solutions:
$$1 + 2 + 4 + 5 + \cdots + 28 + 29 + 31 =$$
$$1 + 2 + 3 + 4 + 5 \ldots + 28 + 29 + 30 + 31 - 3 - 6 - 9 - \cdots - 30 =$$
$$1 + 2 + 3 + 4 + 5 \ldots + 28 + 29 + 30 + 31 - 3(1 + 2 + \cdots + 10) =$$

Note:

62

$$1 + 2 + 3 + \cdots + n = \frac{n(n+1)}{2}$$

Hence,

$$1 + 2 + 3 + 4 + 5 \ldots + 28 + 29 + 30 + 31 - 3(1 + 2 + \cdots + 10) =$$

$$\frac{31(31+1)}{2} - 3\frac{10(10+1)}{2} =$$

$$31 \times 16 - 15 \times 11 =$$

$$496 - 165 =$$

$$331$$

Ex 47 (from RMC 2017)

Consider the positive real numbers x, y, z, with the property that:

$$xy = \frac{z - x + 1}{y} = \frac{z + 1}{2}$$

$$xy = \frac{z - x + 1}{y}$$

So,

$$z = xy^2 + x - 1$$

$$xy = \frac{z + 1}{2}$$

So,

$$z = 2xy - 1$$

We get,

$$xy^2 + x - 1 = 2xy - 1$$

$$x(y^2 - 2y + 1) = 0$$

63

$$x(y-1)^2 = 0$$

x is a positive number, so, $y = 1$

$$xy = \frac{z+1}{2}$$

Hence,

$$x(1) = \frac{z+(y)}{2}$$

$$x = \frac{z+y}{2}$$

So, x is the average of y and z

Ex 48 (from PMO 2022)

Numbers divisible by 25 ends with either 00,25,50,75.

00 cannot happen **be**cause it is repetition 0 also 75 can never come because 7 is not available

<u>1st case number ends with 25</u>

$$\_\ \_\ \_\ 2\ 5$$

For the first digit, because this is a five-digit numbers, you have 3 choices (1,3,4)

For the second digit, you have 3 choices (0 can be in the list), and 2 choices for the third digit.

In total:

$$3 \times 3 \times 2 = 18$$

18 solutions

<u>2nd case number ends with 50</u>

$$\underline{\;5\;0}$$

For the first digit, you have 4 choices (1,2,3,4), for the second digit, you have 2 choices, and 2 choices for the third digit.

In total:

$$4 \times 3 \times 2 = 24$$

24 solutions

Total 5 digits numbers are $18 + 24 = 42$

Answer 3

Ex 49 (from CMO 2023)

Let $abcd$ be Aroon's PIN with

$$0 \leq a \leq 9$$
$$0 \leq b \leq 9$$
$$0 \leq c \leq 9$$
$$0 \leq d \leq 9$$

Aroon's PIN is $1000a + bcd$

"When the first digit (reading from the left) is moved to the end of the PIN, the resulting integer is 6 less than 3 times", hence:

$$3(1000a + bcd) - 6 = 10bcd + a$$
$$2999a = 7bcd + 6$$

Because $bcd < 1000$,

$$7bcd + 6 < 7006$$

$$2999a < 7006$$

$$a < 7006/2999 = 2.336$$

So, a is 1 or 2.

<u>If $a = 1$:</u>

Then

$$2999 = 7bcd + 6$$

$$2993 = 7bcd$$

$$bcd = \frac{2993}{7} = 427.571 \ldots$$

bcd is not a whole number

<u>If $a = 2$:</u>

Then

$$2999 \times 2 = 7bcd + 6$$

$$5998 - 6 = 7bcd$$

$$bcd = \frac{5992}{7} = 856$$

bcd is 856

Hence Aroon's PIN is 2856

Ex 50 (from AIME 2022)

Adults made up $\frac{5}{12}$ of the crowd of people at a concert. After a bus carrying 50 more people arrived, adults made up $\frac{11}{25}$ of the people at the concert. Find the minimum number of adults who could have been at the concert after the bus arrived.

Let x be the number of people at the party before the bus arrives, so x is a multiple of 12.

"After a bus carrying 50 more people arrived, adults made up $\frac{11}{25}$ of the people at the concert." So, $x + 50$ is a multiple of 25 (5 × 5).

x is a multiple of 12 and 25, so the minimum number is $12 \times 25 = 300$.

So therefore, after 50 more people arrive, there are

$$300 + 50 = 350$$

people at the concert.

The number of adults is:

$$350 \times \left(\frac{11}{25}\right) = 154$$

www.ingramcontent.com/pod-product-compliance
Ingram Content Group UK Ltd.
Pitfield, Milton Keynes, MK11 3LW, UK
UKHW010626270525
6096UKWH00041B/448